Tamires Cristina Candido Marques

The notion of electric field: Reflections of a future teacher

Tamires Cristina Candido Marques

The notion of electric field: Reflections of a future teacher

The process of teaching physics

Imprint

Any brand names and product names mentioned in this book are subject to trademark, brand or patent protection and are trademarks or registered trademarks of their respective holders. The use of brand names, product names, common names, trade names, product descriptions etc. even without a particular marking in this work is in no way to be construed to mean that such names may be regarded as unrestricted in respect of trademark and brand protection legislation and could thus be used by anyone.

Cover image: www.ingimage.com

This book is a translation from the original published under ISBN 978-3-330-75942-8.

Publisher:
Sciencia Scripts
is a trademark of
Dodo Books Indian Ocean Ltd. and OmniScriptum S.R.L publishing group

120 High Road, East Finchley, London, N2 9ED, United Kingdom
Str. Armeneasca 28/1, office 1, Chisinau MD-2012, Republic of Moldova, Europe
Printed at: see last page
ISBN: 978-620-3-62393-2

CONTENTS

CHAPTER 1 - INTRODUCTION

This work was mainly inspired by my activities with the PIBID project[1] UNESP in Rio Claro. This work was carried out in a public high school (ETEC Prof Armando Bayeux da Silva) in the city of Rio Claro - SP. We discussed a specific and different methodology for preparing and applying Physics lessons, using low-cost experimental materials.

While working on the topic of electricity, inspired by Prof. Norberto Cardoso Ferreira with low-cost experiments for teaching Physics at secondary level, we identified the possibility of discussing and comparing the use of conventional teaching materials (essentially the textbook) and the use of low-cost experiments applied to primary school students, with the use of the experiment called "Electrostatic Vector".

In choosing this experiment, we focused our attention on teaching the concept of the electric field. The emphasis was on the lines of force of the electric field proposed in the 19th century by Michael Faraday.

It is very difficult to teach an abstract concept in the traditional way, because there is a geometric spatial perception involved, as well as the interactions between the charges. These details are not easily visualized by students when using paper, chalk and blackboards, i.e. in a flat view. A three-dimensional visualization of the problem in question is necessary.

The work presented here is divided into the following chapters:

In the second chapter we discuss a methodological proposal used to teach electric fields and we look at the difference between lectures with exercises *and* lectures with experimental materials. We will look at low-cost electricity materials, with a focus on electric field experiments.

In the third chapter, we analyzed two textbooks for teaching high school physics. Our aim was to differentiate and contrast the eminently theoretical

---

1 PIBID was the Institutional Program for Teaching Initiation Scholarships. Promoted by CAPES.

approach to the teaching of electric fields with that discussed in chapter two of this work.

In the fourth chapter we discuss the physical concepts associated with the electric field, electric charges and the invention of the graphical representation of the electric field through the use of lines of force. We will also look at the life and work of Michael Faraday, an English scientist who made significant advances in electromagnetic theory.

In the fifth chapter we present the low-cost experiment called electrostatic vector.

The sixth chapter will deal with the activity carried out with the low-cost experiment - electrostatic vector, in a secondary school. The differences between the textbooks analyzed in chapter three of this work and the development of the electric field course will also be discussed.

In the seventh chapter we discuss the implications and possibilities of the proposal explored in this work.

CHAPTER 2 - PHYSICS TEACHING

The focus of this chapter will be to discuss the difference between traditional methodologies commonly used in the classroom and the experimental approach that can be given to lessons using low-cost materials.

The experimental material was used as a pedagogical strategy for teaching the electric field. The learner must interact with the electrostatic vector and from this interaction develop skills in order to learn the concept according to their cognitive structures.

The main aim of this chapter is to discuss the type of material used and how easy it is to use this type of material to create experiments that can be applied in the classroom. The main focus is on the electrostatic vector, which is a material from "Ferreira and Ramos (2008)" that allows us to explore the fact that an electrified body has an electric field associated with it.

When it comes to low-cost experiments, some teachers have a slightly prejudiced attitude which will also be presented in this chapter.

2.1-Teaching electric fields.

In my experience as a student in primary education and as a Physics undergraduate, both in the Teaching Practice subject and in the activities of the PIBID project, I have found that Physics lessons are generally as traditional as possible. These classes pay little or no attention to how students learn physics. They traditionally focus on presenting mathematical formulas in an attempt to justify that they represent these concepts, which often seem devoid of any logic or meaning to the student. It is important to point out that many of these equations and formulas represent more of an approximation of the real behavior of nature than nature itself.

Electric field lessons are no exception to this pattern, i.e. they are traditional lessons in which the teacher uses "chalk, blackboard and saliva" to try to explain concepts and characteristics such as spherical geometry, the way the force field is represented and the density of lines. Aspects that go beyond descriptive speech. In this methodology, the exercises play the role of instruments that lead the students to use mathematical tools to solve the "problem".

In this context, it is clear that the physics lessons taught in most schools are very mathematical. The focus is not on the physical concept behind a phenomenon, but on solving, often excessively, calculations. As previously discussed, students often see no point in these calculations.

It is necessary to emphasize that lectures should in no way be disregarded or ignored. It is desirable for the student to reach a level of cognition where the concept can be worked out in an abstract way in eminently conceptual lessons.

We emphasize, however, that it is difficult for teachers to express issues such as distance action, symmetry and the intensity of electric fields, for example, using a traditional approach.

We realized that these characteristics could be much better understood if students could "get their hands dirty". Interacting with an experiment is advantageous for the student, as it gives them the chance to understand the points mentioned in a practical way.

In view of this, an alternative way for students to qualitatively improve their learning of concepts such as the electric field is to use experimental materials, as proposed by Kaptisa.

"For a student to understand an experiment, he must carry it out himself, but he will understand it much better if, as well as carrying out the experiment, he builds the instruments for it. " (KAPTISA,1985).

In this teaching process, the experimental material is used to help learn the concept, and not just "at the point of a pencil", solving math that is supposed to be included in exams or entrance exams.

Some teachers believe that learning is a process that doesn't require experiments, intuition, games and creativity, or simply anything that is different from a traditional lesson, or what we call play.

We realize, however, that the act of curiosity, the desire to handle and the students' interests are often awakened through playfulness. In this way, physics can be taught through processes that are different from traditional lessons. We emphasize that working with low-cost experimental material is just one example of an "alternative" method that can provide this playfulness.

Who says that this type of teaching methodology, the traditional classroom, accurately teaches students concepts (whether in Physics or other areas of knowledge)? Students are often "threatened" by teachers, principals and other members of the school and society itself, who often say that "This subject will be on the exam" or "You're stupid because you don't get high marks". It is at this point that the coerced student may not acquire the knowledge for that particular content. With this kind of didactic attitude, only the problems of the moment appear to be solved, but future problems such as disgust with studying and school are disregarded by teachers.

For Piaget (1983), each person interacts with knowledge and carries out internal operations to build it according to the cognitive structures of each individual. In this way, we can deduce that the construction of knowledge is not reduced to a single type of teaching methodology, but rather to a set of activities that favor learning.

2.2-Low-cost experimental materials.

According to Ramos (1990), low-cost experimental materials are materials that have the technical and pedagogical performance required for a given assembly, but which are also inexpensive compared to other options. Also, the choice of low-cost material is not only due to the fact that it is financially economical, but also due to its simplified construction procedure. Using this type of material allows the student or teacher to explore aspects beyond the construction itself, making changes, adaptations and inventions.

"Ferreira and Ramos (2008)" in "Cadernos de Instrumentaçâo para o Ensino de Fisica" describe various electrostatics experiments. These include the materials used, how to set up the experiment and conceptual indications of physics, which sometimes involve calculations. This shows us that the materials, although simple and easy to construct, express the physics content, highlighting its experimental and didactic aspects.

The choice made to write this paper, which is the use of the experiment called electrostatic vector[2] , was made with students in mind. The concept of the electric field is often left aside and starts to be worked on as an "extra complex formula", in other words, the vast majority of students graduate from high school without properly mastering this concept.

2.2.1- *Teachers' attitudes towards low-cost experiments.*

Observing Physics classes, we realized that no experimental activities are carried out, because teachers claim that it is laborious and requires extra classroom time to prepare and confirm the results. They also believe that they need a space to carry out the experiment, i.e. a teaching laboratory, so that they can carry out these activities with their students. The teachers' lack of familiarity with low-cost experiments leads them to indicate that they need very specific materials that are difficult to access.

Of these "problems" pointed out by a teacher, the one that is most evident in the day-to-day running of a school is the lack of teaching laboratories. This situation presents itself as a justification for a teacher to "teach" students merely mathematical formulas, based solely and exclusively on textbooks.

During my training activities, whether as a graduate student or in the PIBID project, which took place in public schools in the city of Rio Claro, I noticed that the laboratory had become a dumping ground for old textbooks and boxes. However, when I had to use experimental materials for the electric field course,

---

2 The construction of this experiment is explained in detail in Appendix A. It can therefore be reproduced.

I was able to carry out the activity in an ordinary classroom with just tables, chairs and a small blackboard. In other words, it was possible to work in the classroom itself, without this being an obstacle to fulfilling the role of teaching the concept of the electric field in a different way to the traditional one.

The "excuses" for not using experiments therefore don't entirely hold water. I believe that in order to use these teaching resources, it is essential that teachers understand the role of experiments in the construction of scientific knowledge. Teachers can stimulate students' curiosity and to do this they must adopt resources and methodologies that are different from the traditional ones. An appropriate methodology would be to educate students scientifically and not just mathematically.

However, in order for the use of experimental material to be completely effective, teachers need to change their own concept of teaching, as the use of resources alone will not be enough for students to learn.

In order to assess students, it is not only necessary to use a test, but mainly to monitor the students throughout the activity and if they fail to achieve the pre-established goals, the teacher is obliged to give them new tools so that they can develop and learn about a particular subject, but it is clear that this obligation has not been satisfactorily observed.

CHAPTER 3 - TEXTBOOKS AND THE ELECTRIC FIELD

Two primary school textbooks, specifically on the topic of electric fields for secondary schools, were analyzed. The aim of this analysis was to compare the use of the textbooks with the activity carried out with the electrostatic vector experiment.

The first book analyzed was written by José Luiz Sampaio and Caio Sérgio Calçada (2005). It is a single volume book from the 2nd edition[a] , which covers the following subjects: Mechanics, Thermology, Optics, Waves, Electricity, Magnetism and Modern Physics.

The second book is by the authors: "Antônio Màximo Ribeiro da Luz and Beatriz Alvarenga Alvares (2005)". It is a collection of three volumes from the 6th edition. The first deals with Mechanics. The second covers thermology, optics and waves. The third deals with Electricity, Magnetism and Modern Physics.

These books were chosen because they are current textbooks and widely used in many middle schools.

3.1-Sampaio and Calçada.

As mentioned above, this textbook is presented in a single volume for all three years of secondary school, covering subjects from mechanics to electromagnetism. However, the content analyzed in this textbook corresponds to content specifically covered in the 3rd year.

The book contains 472 pages and has chapters that present: theory, solved exercises, proposed exercises and, at the end of the volume, answers to the applied exercises, as well as additional exercises focusing on university entrance exams and Enem.

Table 1 shows the content applied to secondary schools and the number of pages in the book by Sampaio and Calçada (2005).

The content that will be analyzed is within the fifth topic: Electricity and Magnetism.

Table 1: Division of content with the respective number of pages for each specific subject.

Content discussed	Number of pages
Mechanics	162
Thermology	40
Optics	40
Waves	20
Electricity and Magnetism	126
Modern Physics	33

3.1.1- *Critical analysis of the electric field chapters.*

The concept of the electric field is discussed in two chapters containing eight pages. Four in each chapter.

The first chapter (chapter 49 of the book) begins with an example from the students' everyday lives: the observation that when you bring your arm close to the TV screen, your hair bristles, showing that the electric charges in the TV generate an electric field.

However, this observation is only valid for some types of TV that use cathode ray tubes. Furthermore, the physical processes involved in this example are not explained. The authors simply say that the effect is generated by electrical charges.

They then introduce the subject of proof charge, electric field strength, direction and direction. But they don't explain each of these in detail, meaning that the vectorial characteristic of the electric field is not properly explored.

They show the directions of the lines of force of positive and negative charges and, in order to solve these electric field problems, they define the formula that relates the electric field, electrostatic force and proof charge. The unit of measurement corresponding to this formula is then presented.

They demonstrate mathematically how to deduce the formula for the modulus of the electric field at any point in two-dimensional space.

They describe that the resulting electric field is given by the vector sum, "playing" the cosine law formula, which contains errors. The formula as it appears in the book is: $E_{res} = E1 + E2 + 2. E_1. E2. Cos\theta$, where θ is the angle between electric field vectors 1 and 2. We can see that the squares in the first two terms $E1$ and E_2 are missing for the equation to be correct. We are also missing a square root that encompasses the right-hand side of the equation.

In another chapter, the authors once again use mathematics to introduce sums of electric field vectors.

At the end of the chapter they suggest simple experiments for students to do at home and visualize the effects. However, these experiments do not correspond to the three subchapters analyzed, but rather to the other two that were not explored.

These are solved exercises for calculating the value of the electric field using vector summation.

To sum up, it can be seen that when using this textbook as the basis for a lesson, chalk and blackboard are the main resources, without any experimental material or dialog with the students. In this way, the student's interaction with knowledge is limited. Students are expected to be able to manipulate and execute the mathematics presented, with the aim of answering the proposed exercises numerically.

The language used by these authors is easily accessible to those who already have a background in Physics. In other words, the book contains phrases and concepts in a language that does not appeal to students interacting with this material for the first time.

3.2-Antônio Mâximo and Beatriz Alvarenga.

This collection has 3 volumes. I analyzed volume 3 of the textbook, which is aimed only at the 3rd year of secondary school, as it covers Electricity, Magnetism and Modern Physics. This book contains 440 pages with chapters on theory, proposed exercises and exercises for university entrance exams and Enem. In the case of the teacher's book, there are a further 128 pages of teaching advice.

Table 2 shows the content breakdown of this textbook.

Table 2: Breakdown of content with the respective number of pages for each subject in Màximo and Alvarenga's textbook.

Content	Number of pages
Electricity and Magnetism	318
Modern Physics	62

3.2.1-   Critical analysis of the electric field chapter.

This chapter on electric fields is subdivided into five subchapters. These deal respectively with: The concept of electric field. Electric field created by point charges. Lines of force. The behavior of an electrified conductor. Dielectric strength - tip power.

The first three subchapters have been analyzed, since only these relate to the content of this work.

The chapter begins with the concept of the electric field, through a theoretical explanation, as well as figures that represent this phenomenon. It is described that the electric field is a vector, i.e. it has intensity, direction and sense, which is represented conceptually, through figures and definitions through formulas. The book also includes solved examples, but these are numerical and contain little theoretical concept. The exercises proposed for students to "learn" are called retention exercises, most of which are numerical, even though they contain some illustrative figures to help with reasoning.

Starting from the premise that the electric charge generating the electric field is a point charge. The authors use the mathematical definition of this electric field and describe a new formula using Coulomb's Law.

Comments are made and graphs analyzed, where the electric field decreases with the square of the distance between the charges.

They also describe how to calculate the electric field of several point charges by summing vectors. To make it easier to understand, a drawing has been made of several charges with their respective electric field vectors.

In the next subchapter, the authors give a historical overview of who introduced the concept of lines of force. They tell about the life of Michael Faraday, and with illustrative figures they talk about the lines of force of an isolated positive charge and an isolated negative charge, which are respectively lines of approach and distance. Through comments and images, they show the electric field represented by the lines of force of two charges of opposite signs and another of equal signs.

Even though it contains mathematical formulas and definitions that are often not fully explained, a student who doesn't understand physics will be able to understand certain concepts through this book. The language is not only treated from the point of view of "physics" (with jargon and expressions typical of physicists). Any extra questions can be answered by the teacher.

As in the previous book analyzed, this one also uses a "chalk and board" methodology, which does not favor the use of experimental material in the classroom. The use of experimental materials could help visualize the lines of force of the electric field and not merely decorate the definition proposed by Faraday.

CHAPTER 4 - ELECTROSTATICS CONCEPTS

"Electrostatics: Branch of Physics that investigates the properties and behavior of electric fields, electric charges or stationary charge sources" (RODITI, 2005, p.76).

Initially, this chapter will deal with non-accelerated electric charges. The concepts of electric field and lines of force will then be presented, as well as a historical discussion, the development of this concept and the life and work of Michael Faraday.

4.1-Principles of Electrostatics.

The study of electrostatics is considered to have begun around 600 years BC "(YOUNG, 2010)". It is usually cited as the beginning of such studies, the discovery that when rubbing wool and a "stone" of amber, a previously unknown interaction occurred between these elements - the fact that they attracted small objects. This discovery is attributed to Thales of Miletus. Today we believe that the attraction of amber to other objects is due to the presence of electrical charges, an intrinsic property of matter.

"Electric charge is an intrinsic property of the fundamental particles of which matter is made; in other words, it is a property associated with the very existence of particles" (HALLIDAY, 2012, p.1).

There are two types of electric charge, negative charges (conventionally associated with electrons) and positive charges (conventionally associated with protons). In an electrically neutral body, the amount of positive electric charge is equal to the amount of negative electric charge.

When two different materials rub together, they become electrified, both with opposite charges, because when they rub together, materials that tend to receive electrons attract electrons from the other type of material that tends to donate them. To identify who is negatively or positively charged, you can use a table called the triboelectric series. It was constructed experimentally considering different materials and their electronegativity.

As electrons are the mobile charge carriers in solid bodies (due to their reduced mass and weak interaction with the nucleus for some atoms), we treat the electrification of a body by considering the excess or lack of electrons in relation to protons. If a body has an excess of negative charges, it is said to be negatively charged. If it has a *deficit of* negative charges, it is said to be positively charged. Electrically charged bodies interact with each other. We treat this interaction as electrical forces, which can be either a force of attraction or a force of repulsion. For bodies charged with the same electric charge, we consider there to be a force of repulsion. If the charged bodies have opposite charges, we consider there to be a force of attraction. This force of repulsion or attraction is called the electrostatic force.

The empirical law that allows us to calculate the force of attraction or repulsion is Coulomb's Law. Below is the formulation of this law.

"The force between two point charges is exerted along the line of the charges. It varies with the inverse square of the distance separating the charges and is proportional to the product of the charges. The force is repulsive if the charges have the same sign and attractive if they have opposite signs" (TIPLER, 2009, p.6).

Written mathematically, the electrostatic force tells us:

$$\vec{F} = \frac{k.q_1.q_2}{r^2}\,(\hat{r}). \tag{1}$$

Where F is the electrostatic force, q_1 and q_2 are the intensity of the electric charges, is the distance between the charges, k is a proportionality constant, called Coulomb's constant and is the direction in which this force points. In the case of point charges, this force is radial, i.e. from a central point the force extends inwards or outwards. It is important to note that these forces act at a distance, i.e. without contact between the electric charges. If we brought two charges close enough to touch, we would have $r=0$ and an infinitely large force as a result.

We can also consider the electrical behavior of matter based on a charge, so instead of considering forces, we start to consider the influence of the charge on the surrounding environment. This change of reference translates into the physics concept of the electric field.

Making an analogy, the Earth and a massive body attract each other because there is a gravitational field. In the case of electric charges, we also treat them as a force field, but unlike the gravitational field (which always tends to bring two masses closer together), they can be attractive or repulsive.

Although always present, the electric field of a charge can only be detected by approaching a second charge.

Mathematically, the electric field felt by the second charge is:

$$\vec{E} = \frac{\vec{F}}{q}. \qquad (2)$$

Where is the electric field produced by the first charge, is the electric force exerted and q is the second charge that feels the electric field, placed near the original electric charge.

Figure 3 shows the electric field of a positive charge and Figure 4 shows the electric field of a negative charge (both isolated from other charges).

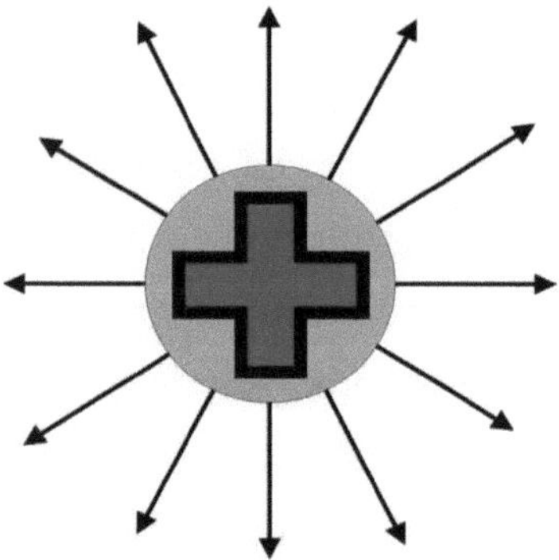

Fig 3: Positive charge with distance lines.

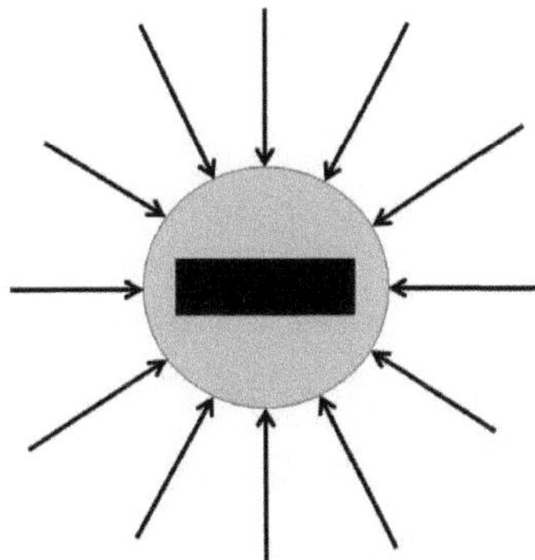

Fig 4: Negative charge with approach lines.

The electric field is represented by a vector field. By drawing a set of lines, you get a representation of this vector field at each point in space. Analyzing the density of these lines shows that it is proportional to the magnitude of the field, i.e. the more lines together per cubic meter, the more intense the electric field, while separate lines have a lower intensity.

Representation by lines of force has advantages because it creates an image of the effect of an electric charge on its surroundings, a kind of "aura", so that it is not just an abstract concept for those trying to learn the topic.

It was Michael Faraday who introduced the concept of lines of force, which helped visualize the electric field created by these positive and negative charges.

The lines representing the electric field are distance lines and approximation lines for positive and negative charges, respectively. To understand this representation, we must assume that a test charge q, or also known as a test charge, is always positive and that the value of its electric field is negligible when compared to the electric fields of other positive or negative charges.

In order to identify the lines of force of the positive charge, the test charge q must be brought closer to the positive charge Q_+. When the electric field of the charge Q_+ "identifies" the other charge q around it, this charge Q_+ exerts a force on the test charge in order to push it away, due to the fact that the charges have the same sign. To represent the electric field of this positive charge, Faraday drew lines coming out of the positive electric charge Q_+ itself, representing the repulsion force (figure 3).

For the representation of the lines of force of negative charges, it was assumed that when the test charge q approaches the negative charge Q_-, the electric field of this negative charge "identifies" the test charge and the latter is then pulled closer to Q_-, due to the fact that they have opposite signs. To do this, Faraday drew lines entering the negative charge Q_-, symbolizing the "pull" that the negative charge infers on the test charge, which describes the attraction between the charges (figure 4).

4.1.1 Equipotential surfaces.

Equipotential surfaces are defined as curves in space that contain the same electric potential. The relationship between field lines and equipotential surfaces is that the electric field is the gradient of the electric potential.

"Electric potential: defined as the work required to move a positive electric charge from infinity, where the potential is zero, to a given point in the electric field" (RODITI, 2005, p.181).

"This gradient points in the direction in which the potential changes most rapidly, and is therefore perpendicular to equipotential surfaces. If the electric field were not perpendicular, there would be a component on this surface and consequently it would not be an equipotential surface" (FEYNMAN, 2008, p. 4-13).

"Gradient: The rate of directional variation of certain quantities characteristic of a medium, such as atmospheric pressure, temperature, etc" (RODITI, 2005, p.110).

To construct equipotential surfaces of one or two point charges, first draw the lines of force of the electric field and then draw a perpendicular at each point on these lines.

In the case of an isolated point charge, the equipotential surfaces are centered on the charge, thus forming either a three-dimensional sphere as shown in figure 5 or a circle in the plane. Figure 5 shows this case for a positive charge, but we can extend it to a negative charge, which differs only in the directions of the lines of force.

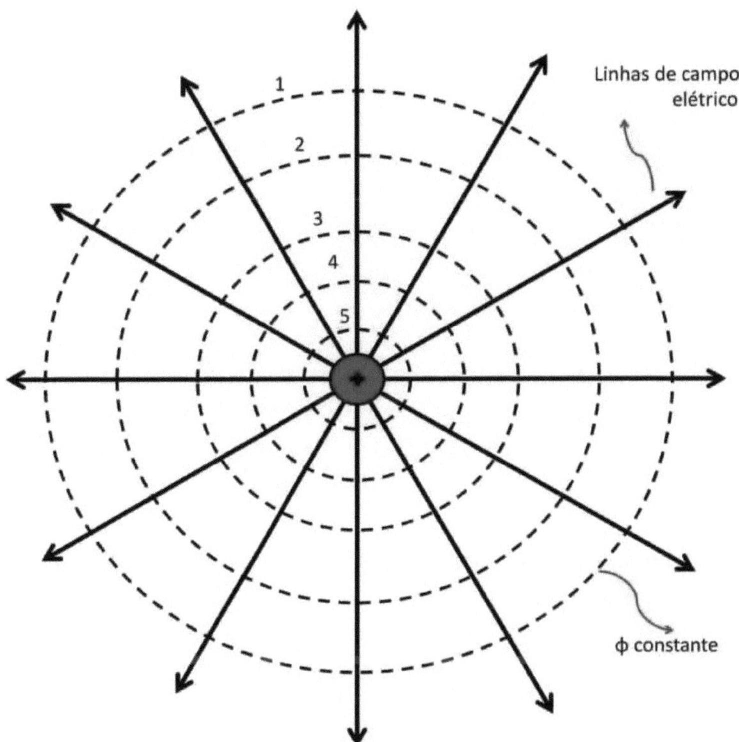

Fig.5- Equipotential surfaces and lines of force of a positive charge. Image based on the book Lições de Fisica "(FEYNMAN, 2008)".

In the case of two electric charges of opposite sign, the electric field is the superposition of the individual electric fields. The equipotential surface is neither a three-dimensional sphere nor a circle in the plane. Figure 6 shows us what equipotential surfaces look like in the case of two charges.

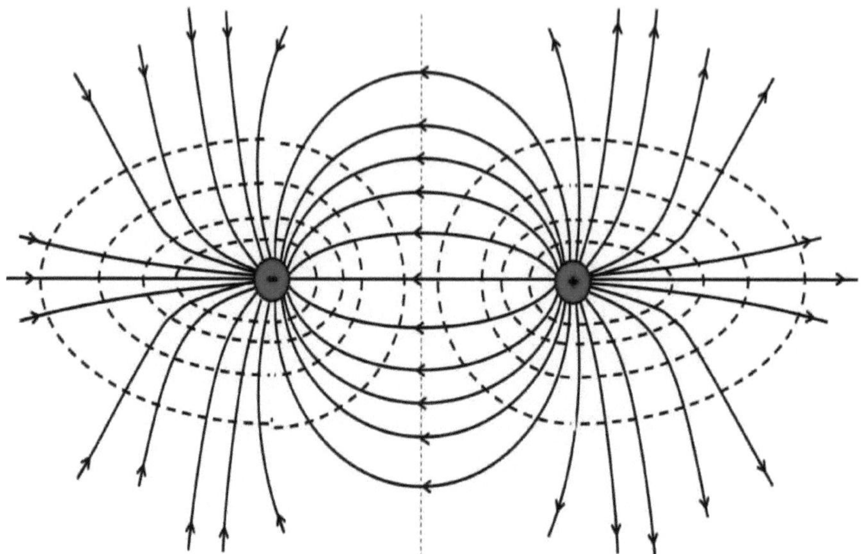

Fig. 6- Equipotential surfaces and lines of force for the case of two charges of opposite signs. Image based on the book Lições de Fisica "(FEYNMAN, 2008)".

To represent the intensity of the electric field vector, the "density" of lines will be used, as mentioned above. In the case of a point charge, the density of these lines decreases with the inverse of the distance squared. However, the area of the surfaces perpendicular to the lines at a radius of r increases with r^2 , i.e. the closer to the central charge, the closer the equipotential lines, and the further away from the central charge, the greater the spacing between these lines, thus decreasing the density of lines with distance.

4.2-Michael Faraday - a brief biography

Fig.7 - Faraday as a young man

Michael Faraday was born on September 22, 1791 in Newington Butts, Surrey - London. Son of James Faraday and Margaret Hastwell. His family came from humble beginnings. His father, realizing that his job as a blacksmith didn't meet the family's basic needs, moved with the family to London in order to get a better job. The situation worsened when his father died in 1809.

With his father gone, Faraday had to start working at the age of 13 to help make ends meet, delivering newspapers for Mr. Riebau. When he turned 14, Mr. Riebau hired him and taught Faraday the art of bookbinding. He worked in this field for the next seven years. As a result of getting this job, Faraday developed extraordinary manual dexterity, which later helped in his experimental research.

With books all around him, Faraday's mind, despite having little knowledge of mathematics and reading, was stimulated and he began to make notes in a notebook in which he included quotes and observations that might be interesting. He attended lectures and took notes. Faraday's passion for science was awakened when he was binding an encyclopedia *(Encyclopaedia Britannica)* which contained the entry "Electricity".

In 1810 he attended his first class at the City Philosophical Society, led by John Tatum. They met at the leader's house and he would give a lecture on some scientific topic and open his library to the members. It was through this society that Faraday discovered a book that sparked his interest in chemistry.

He won a ticket to attend chemist Humphry Davy's lecture at the *Royal Institution*. He took notes and wrote them down to share with his friends in the Society.

In 1812, he finished his apprenticeship and devoted himself solely to bookbinding. Faraday was recommended as a copyist to Davy, and he took advantage of this opportunity to send his carefully bound notes to the lecturer asking for a job, but it wasn't until the following year that he was asked to fill the vacancy of an assistant who had been dismissed, at the *Royal Institution.*

In October 1813, he traveled with Davy to France, Italy and Switzerland, where he met several important scientists such as Alessandro Volta and Joseph Gay-Lussac. It was during this period that he learned more about a scientific career.

In 1821 he married Sarah Barnad. In the same year, he developed various works in chemistry (studies on chlorine, gas diffusion, liquefaction, among others).

In 1824, thanks to his work in chemistry, he became a member of the *Royal Society.* The following year he became its director.

His research into electrochemistry offered a new view of electrostatics. In the 1780s, Coulomb discovered that the electric force acted at a distance, in other words, he solved the question of action at a distance, but Faraday wanted experimental confirmation of his point of view.

In 1839 Faraday had a nervous breakdown from which he never recovered. He spent the next five years studying nothing about electricity and magnetism, but continued at the *Royal Institution*, dedicating himself to research that didn't require so much of his commitment.

On August 6, 1845, William Thomson (Lord Kelvin) sent a letter to Faraday telling him of his success with the mathematical approach to the concept of lines of force. He spoke of his experiments and Faraday reproduced them and thus managed to detect the phenomenon he had been looking for since the 1820s.

His mental faculties declined until, in 1850, Faraday withdrew from social gatherings and tried to concentrate on the *Royal Institution.*

In 1862 he retired from the *Royal Institution* and moved to a house that Queen Victoria offered him at Hampton Court.

He died on August 25, 1867 in Hampton Court - London.

4.3-Faraday and the concept of lines of force.

Around 1821, Ostered obtained the first experimental results on electromagnetic induction. He demonstrated that electrical and magnetic phenomena were not independent of each other. The experiment he used to prove this phenomenon was that of an electric current passing through a conductor, causing the magnetic needle of a compass to move.

With this discovery, Faraday, feeling curious, began to study and reproduce experiments carried out in the past and those of the time.

In his studies, Faraday began to notice that whatever was the cause of magnetism was the manifestation of something that surrounded the magnet, in other words, it was in the field and not in the magnet (body). This manifestation was from that time on known as the magnetic field.

With his experimental results, Faraday ended up making analogies between the magnetic field and the electric and gravitational fields.

In order to better understand the concept of electric, magnetic and gravitational fields, he created the representation of lines of force.

These lines differ in some respects. For example: electric and gravitational fields have straight lines of force (in the case of the electric field, considering a point charge). In the case of the magnetic field, they are curved lines, which leave one pole and enter another magnetic pole. He also realized that electric field lines differ from gravitational field lines, because while electric field lines are approaching and moving away, gravitational field lines are only approaching.

However, it wasn't until the 1850s and 1860s that these concepts of electric and magnetic fields were readjusted and written down mathematically by James Clerck Maxwell.

CHAPTER 5 - ELECTROSTATIC VECTOR

Electrostatic vector is the name given to a low-cost experiment developed by Norberto Cardoso Ferreira "(FERREIRA, 2008)". The aim is to show that a charged body has an electric field associated with it.

5.1-Ways of charging the electrostatic vector

There are two possible ways of electrifying this electrostatic vector: by contact or by induction.

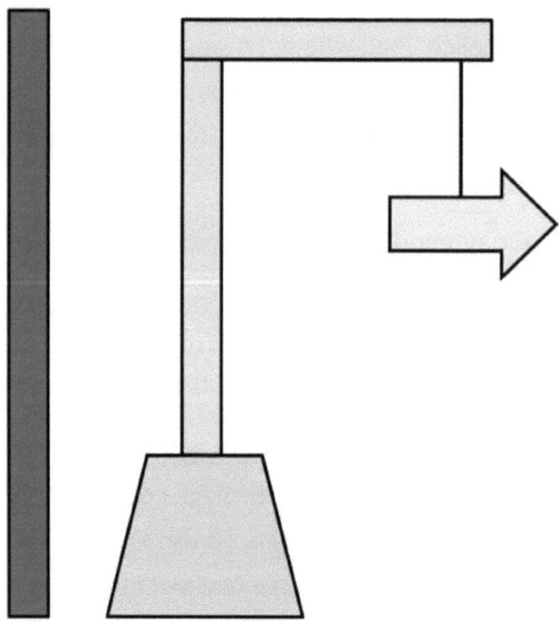

Fig.8- Schematic drawing of the electrostatic vector (straw, plaster base, bendable straw, cardboard, nylon thread - these are used in the experimental set-up, described in the appendix).

Initially, the straw and the toilet paper should be rubbed together. Both are electrically neutral, i.e. the amount of negative and positive charges is equal in each of them.

To start the process of electrification by friction, the straw is rubbed against the toilet paper. There will be a rearrangement of charges and due to the electronegativity of each material, the soft drink straw "steals" electrons from the paper, thus becoming negatively charged and leaving the toilet paper positively electrified.

Figure 9 shows a schematic of the rearrangement of charges after friction. Charges are conserved before and after the materials are rubbed. In other words, the initial quantity of charges will be the same after friction, but not on the same body, but on both, if the system is isolated (without contact with a possible external agent that could influence the rearrangement of charges).

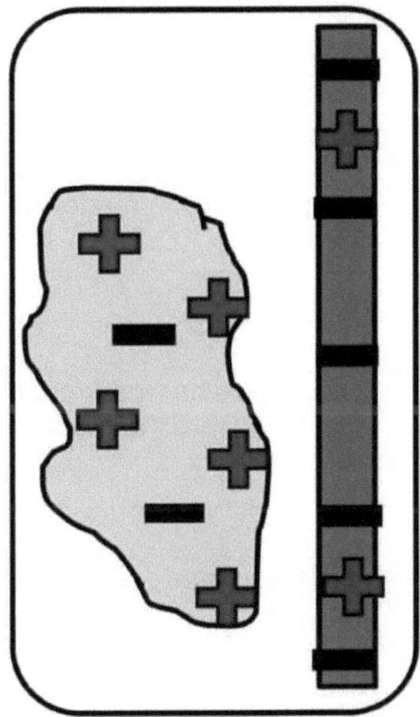

Fig.9-"Visualization" of electric charges after friction. Toilet paper on the left and straw on the right.

To find out which type of material "steals" electrons for itself, leaving the other material with a *deficit,* check the table below, which is based on the triboelectric series found in both high school and undergraduate physics textbooks.

The electronegativity of the material in this table follows a simple rule: the further "up" the material is, the greater its tendency to lose electrons, making it electrically positive.

Table 3: Rubbed materials and their tendency to receive and donate electrons. Based on the

LOAD	MATERIALS	OBESERVATIONS
Material with a greater tendency to become positively charged.	Dry human skin	Great tendency to donate electrons and become highly positive.
	Leather	
	Rabbit fur	It is widely used in friction electrification.
	Glass	The glass of your TV screen becomes electrified and attracts dust.
	Human hair	Combing the hair is a good technique for obtaining a moderate load.
	Nylon	
	Lâ	
	Lead	Lead retains as much static electricity as cat fur.
	Cat fur	
	Silk	
	Aluminum	It lets some electrons escape.
	Paper	
Neutral	Cotton	The best of the "non-static" clothes.
Neutral	Steel	It is not used for friction electrification.
	Madeira	It attracts some electrons, but is almost neutral.
	Amber	
	Hard rubber	Some combs are made of hard rubber.
	Nickel and copper	Copper brushes are used in the Wimshurst

		electrostatic generator .[3]
	Brass and silver	
	Gold and platinum	These metals attract electrons almost as much as polyester
	Polyester	Polyester clothing is hungry for electrons.
	Styrofoam	Often used in packaging. Good for experiments.
	PVC film	
	Polyurethane	
	Polyethylene	
	PVC	Polyvinyl chloride has a great tendency to receive electrons.
Material with a greater tendency to become negatively charged.	Teflon	Highest tendency to receive electrons among all those on this list.

To represent the lines of force of electric charges, we must bring positively and negatively charged bodies close to a test charge, which will always be positive.

Each material in the experiment will play a role in demonstrating the concept of the electric field. The electrostatic vector will play the role of the test charge. The drinking straw, on the other hand, will have the function of representing the positive charge or the negative charge, depending on the experimental approach of interest to the teacher. In the course presented, both approaches were used.

5.1.1- Contact electrification

To observe the phenomenon of contact electrification. Make the negatively charged straw touch the electrostatic vector. It will happen that some of the negative charges from the straw will flow into the vector.

By placing the electrified straw on a plaster base and rotating the electrostatic

---

vector around the straw, we can see that the electrostatic vector will always point in the opposite direction to the straw (figure 10).

The physical explanation stems from the fact that there are equal charges on both bodies and so there is a repulsive force between them, causing the vector to rotate by 180°.

Figure 10 shows the mapping of the negatively charged vector around the straw, which is also negatively charged.

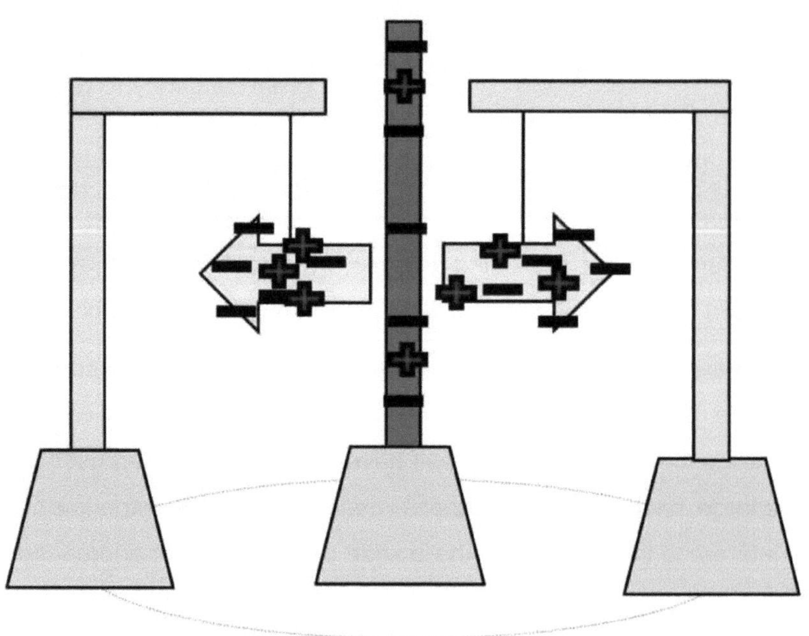

Fig.10 - Schematic drawing of the mapping of bodies with the same load.

Why, instead of just moving away from the straw, did the electrostatic vector turn 180°, pointing in the opposite direction?

The answer to this question lies in the shape of this electrostatic vector - an arrow.

In physics, there is a concept which describes that electric charges have preferences for points, which we call "*The Power of Points*". The application of this concept is linked to our daily lives. One example is the lightning rod, which attracts electrical discharges from clouds (lightning) so that no one gets hurt.

The electric charges then behave in such a way that they feel an "attraction" to the tips, clumping together. As we know, it is the negative charges that move around inside matter, so they move from the back of the vector's body to the front, where the tip is.

As there will be more negative charges at the tip of this vector, the arrow of the vector tends to turn 180°, pointing in the opposite direction to the straw. This rotation represents the force of repulsion between the negative charges belonging to the two bodies.

However, there is a conceptual peculiarity to this experiment. The direction of the vector indicates that the electric field lines are moving away from the negatively charged straw. This result contradicts the theory presented in Chapter 4. This is because we used a negative proof charge (the arrow in the vector). If the arrow were positively charged, as presented in the theory, it would be pointing towards the straw, correctly representing the field lines.

Let's make an analogy, assuming that the bodies, both the electrostatic vector and the straw, are positively charged (this is because the test charge must always be positive). We can then verify what was proposed in the 19th century: that the lines of force of the electric field of a positive charge must always be lines that originate in the body itself and have a direction away from it.

This analogy is made because when we rub toilet paper with a drinking straw, the straw doesn't "lose" electrons, but acquires them due to its electronegativity, which makes it negatively charged.

For the proof charge to be positive, we can "mentally invert" the charges. In this way, we can consider negative charges as positive and vice versa. In the schematic drawing (figure 10), this will represent the same condition, i.e. the vector will point "outwards" from the straw, because in both cases the bodies will be charged with the same electric charge.

5.1.2- Electrification by induction

To electrify the electrostatic vector by induction, we must bring the negatively electrified straw close to the electrostatic vector. Then touch the arrow of the vector with your finger. We then remove the finger and then the straw (inductor). This will ensure that the vector is charged by induction, i.e. that both the straw and the vector are electrified with opposite charges.

When we move the negatively charged straw closer to the vector, which is initially neutral, there will be a separation of charges within the vector. This separation of charges is called polarization, and causes the electrons in the vector and the straw to move as far away from each other as possible. When we place our finger on the body of the vector, some of the negative charges there flow from it to our hand, due to the fact that there is a repulsion force between the charges of the straw and the vector, thus leaving the electrostatic vector with a *deficit of* electrons (positively charged).

After electrifying the vector by induction, we should place the straw on a plaster base and rotate the electrostatic vector around this electrified straw. We will be mapping the electric field lines of negative charges. Represented by the negatively electrified straw and the positively electrostatic vector. Note that in this case it is not necessary to "reverse" the charges, since the test charge (electrostatic vector) is already positively charged.

In this case, we obtain the directions of the lines of force of a negative charge, which are lines that appear at infinity and extend to the charge. These lines are therefore called approximation lines.

Figure 11 shows a schematic drawing of the mapping of oppositely charged bodies.

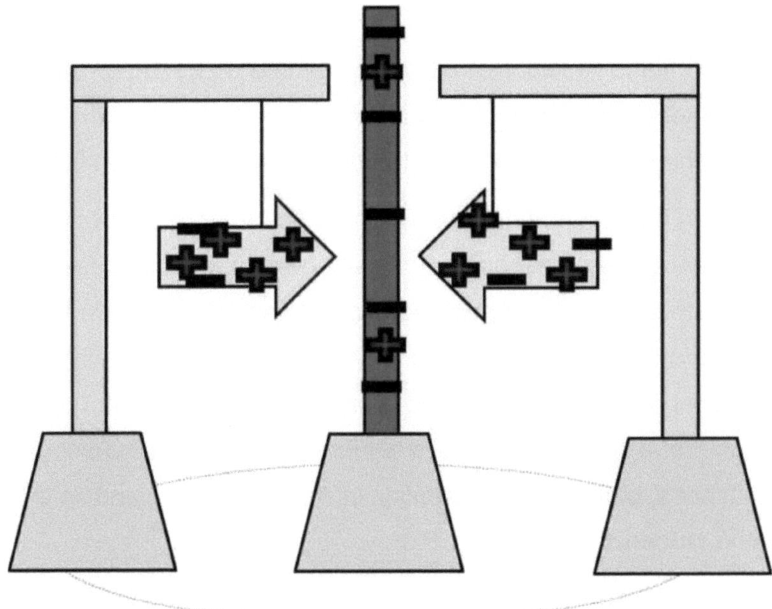

Fig.11 - Schematic drawing of the mapping of oppositely charged bodies.

5.2-Critical analysis of the use of the material.

One of the advantages of using this type of low-cost material is that it is easy for students to handle. They don't run the risk of breaking or damaging the prototype. Generally, experimental teaching materials bought by schools are more expensive and for this reason they are kept in cupboards and not used because the students or any other member of the school could break them, causing damage to the school. In short, the materials present in the school are kept as relics.

This type of low-cost experiment helps students to build their own knowledge in a playful way. The students are not prevented from handling the experiment and even making any changes they see fit.

The experiment is an aid to teaching and learning, because it is not an experiment in which the student already knows what to do. On the contrary, the teacher must play a role in guiding them through the process of learning this subject.

The material used contains limitations that can be overcome by the presence and guidance of a teacher. We found that in fact not everything that the electrostatic vector shows us must be accepted physically (see section 5.1.1).

The example that can be mentioned is that when bodies are charged with the same charge, both negatively and positively they show us the same case, i.e. the direction of the arrow will always point "outwards" and physically we know that this concept fits only in the case of positive charges, thus excluding the case for negative charges from this process. This understanding is not trivial and generally students would not be aware of this fact without the teacher's help, i.e. the teacher's presence is essential in the classroom and in the process of building students' learning.

CHAPTER 6 - ELECTRIC FIELD COURSE

Based on my research for the PIBID project, a course was held with the aim of using low-cost experimental material to check the help it provides on the subject of electric fields.

According to my conceptions of physics teaching and methodologies, I proposed a course to the students, where each individual present in this activity would come to know the electric field, interacting with the material since its construction.

6.1-Course concept and proposal.

The course was divided into four stages. Each stage is described below.

- First stage:

The students were asked questions in order to arouse their curiosity and to develop their answers during and at the end of the activity.

o "What is an electric field?"

o "Why is the shape of the vector an arrow?"

o "Couldn't it be a circle?"

At the same stage, all the low-cost materials needed were distributed so that the students themselves could build their own prototype of the electrostatic vector, based on a ready-made mold, and if the teacher needed help, she was ready to help. But in the meantime, I was watching them.

- Second stage:

I introduced theoretical discussions about the types of electrification (contact and induction), which the students would later use in the experimental part. We then discussed the procedure for charging the electrostatic vector in both ways. At all times, the theoretical discussion was carried out through dialogue with the students. The students responded according to how their learning was built up over the course of the conversations.

- Third stage:

The experimental procedure was carried out (see chapter 5).

The electrostatic vector was charged both by contact and by induction and in both cases the directions of the force representing the electric field of positive and negative charges were mapped.

- Fourth stage:

After carrying out the experimental activity, there was a theoretical recap of the concepts in order to listen to each student and assess whether the experiment had a good effect on learning.

And based on the new discussion, the students themselves answered the questions initially proposed.

To finish off the activity, students were given exercises to draw the lines of force, i.e. to fully verify that the experiment is a teacher's aid.

Table 4 summarizes the steps and procedures carried out on this day of the field course.

Table 4: Electric field course steps and procedures

Stages of the course	Procedure
1^a	Motivating the students with the initial questions and building the low-cost experiment.
2^a	Theoretical discussion of electrification
3a	Electrostatic vector electrification and electric field mapping
4a	New theoretical discussion and application of exercises.

Below is the questionnaire administered to the students.

1) Explain the process of vector electrification.

a) By contact.   b) By induction.

2)Based on the experiment you carried out (mapping positive and negative charges), draw the lines that represent the electric field of the charge.

a) Positive   . b) Negative.

3)Assuming two charges, sketch in the electric field lines

a)   a positive charge and a negative charge.

b)   two positive charges.

c)   two negative charges.

4) Give your comments on the electric field course and your suggestions for improvement, if necessary.

6.2- Differences between the textbooks analyzed and the activity developed with the electrostatic vector.

I conclude that the activity with the low-cost experiment gives students a better understanding of the concept of the electric field compared to the books by "Sampaio and Calçada (2005)", "Màximo and Alvarenga (2005)". Because in the activity the students built their own material, handled it and interacted with it. We believe that they have acquired the conditions for building their own knowledge.

The main reason for this conclusion is that the experiment, electrostatic vector, in operation allows students to perceive characteristics that are difficult to see when reading the textbook, such as the spherical geometry of the electric field.

"Piassi (1995)" points out that various reasons influence the teaching of Physics and that current high school textbooks in particular tend to focus more on mathematical formulas than on the concept itself. This is due to the influence of course books, which aim to train students to pass university entrance exams, leaving important scientific knowledge behind.

Science education today suffers from a lack of investment in school infrastructure, poor teacher training and the result of years of influence from textbooks derived from pre-university course books, which have helped to produce a curriculum based on jargon, formulas and definitions that are disconnected from students' training needs and relevant scientific knowledge (PIASSI, 1995, p.2).

In order for the lesson not to become tiresome and uninteresting for the students, it is important that the teacher uses teaching methodologies that help the students build their learning.

In the case of this activity, low-cost experimental material was used, which in turn helped to visualize the spherical geometry of the electric field.

In addition to the experiences I had as a student, I noticed that some teachers use the textbook as just an exercise book, thus disregarding the potential of the textbook. However, let's be clear that I'm not generalizing this factor to all teachers, as there are still those who use the theoretical discussions presented in the textbook.

CHAPTER 7 - FINAL CONSIDERATIONS

Faraday, despite having little mathematical knowledge, had his interest in reading awakened at a young age. He began reading articles and ended up redoing experiments already carried out by previous scientists of the time, in order to analyze and understand what was happening in them. From these, he drew his own conclusions. One of them was the proposition of the concept of lines of force, without any associated mathematical method.

Due to the fact that Faraday only proposed a theoretical concept based on experiments, we did this work and proposed the same situation to 3rd year high school students, but with a low-cost experiment.

In the classroom experiments using low-cost experimental materials, we noticed that the learning was significant. The concept of lines of force, which represent the electric field, was learned without introducing any mathematical formulations. However, the experiment alone does not teach the concept to the students, but the fact that there is a whole - teacher, experiment and the student's interaction with both - means that the students are able to create higher cognitive capacities and, consequently, the student is able to reach the desired knowledge for the given content.

However, we're not saying that a traditional lesson without any kind of experiment can't arouse interest in students so that they learn. Nor does the fact that students are offered an experiment mean that they learn without even knowing how to handle it, how it works and what the experiment is for.

What we have tried to show with this work is that the lesson with the teacher does not cancel out the experiment, nor does the experiment cancel out the presence of the teacher, but both complement each other, so that the teacher can discuss the subject of the experiment and how it works. The experiment aims to show what neither the textbook nor the teacher can do. One example is the spatial perspective of electric field lines of force.

What we realized in the course presented and consequently our conclusion for this work, is that as textbooks are more focused on mathematical formulations, following the methodology of pre-university course handouts. The introduction of the low-cost experiment helped students learn the concept and visualize the electric field. The students were able to handle the material in a way that they understood in order to visualize the lines of force. What we do know is that the experiment does not replace the theoretical explanation given by the teacher or the mathematical representations, which are part of this concept, but it does help in the learning process.

When we say that the experiment doesn't replace the presence of a teacher, it's because not everything that the electrostatic vector shows really represents the concept in Physics. Because of this, the teacher's presence is irreplaceable so that he can explain what is really happening and why what this experiment shows is not always the concept of the electric field. In other words, there is a big leap between the experiment containing its limitations and the fact that it almost entirely represents the content presented.

We use low-cost experiments because the students have to interact with the experiments and even, if necessary, modify them with their own hands, without the fear of damaging them and having to store them again in "perfect condition". Not handling or modifying the experiment is not the learning process that "Kapitza (1985)" proposes.

The aim of setting up one experiment per student is that they can improve the experiment in the future using the concepts they have learned. The dissemination of the experimental activity among the students, based on their own initiative, is also an important aspect.

Other reasons for choosing a low-cost material are that it's easy to find and buy, and especially that it doesn't work at the touch of a button, since children and teenagers are used to dealing with electronic devices, which ends up depriving them of the creativity to build the experiment and even invent a way of playing or experimenting.

In summary, this work has shown how a low-cost experiment can help a teacher without the need for teaching laboratories, without a lot of money when compared to other materials, and with the advantage of being refurbished. However, many students and teachers don't give the necessary value to low-cost experiments, because they believe they are cheap and don't have the beauty of other purchased ones.

It should also be made clear that if a school has expensive experiments and teaching laboratories available and ready to use, it doesn't mean that the students won't have completed their learning at this stage if they use them. However, the teacher must draw the students' attention to the fact that the effects and aesthetics of the equipment are not as important as the fundamental physics involved.

BIBLIOGRAPHICAL REFERENCES

FERREIRA, N. C; RAMOS, E. M. De F. **Cadernos para instrumentaçao para o ensino de Fisica: eletrostâtica 1ª ed.** Rio Claro (SP): UNESP IB, 2008.

FEYNMAN, R. P; LEIGHTON, R. B.; SANDS, M. **Lições de Fisica.** Porto Alegre: Bookman, 2008.

HALLIDAY, D; RESNICK, R. **Fundamentals of Physics, vol.3, 9th ed,** Rio de Janeiro: LTC, 2012.

HEWITT, P. G; **Conceptual Physics.** Porto Alegre: Bookman, 2002.

KAPTISA, P. *Experiment, Theory and Practice: articles and conferences,* Moscow, Ed. Mir, 1985.

LUZ, A; ALVARES, B. **Curso de Fisica, vol. 3, 6th ed,** Sâo Paulo: Scipione, 2005.

PIAGET, J. **Problems of genetic psychology.** In: The

Thinkers. Sâo Paulo, Abril Cultural, Sâo Paulo, 1983.

PIAGET, J. **Epistemologia genètica.** Translated by Alvaro Cabral. Sâo Paulo: WMF Martins Fontes, 2007.

PIASSI, L. P. C. **What Physics to teach in secondary school: elements for reworking the content.** 1995. 208 p. Dissertation (Master's Degree in Science Teaching). Institute of Physics and Faculty of Education, University of São Paulo.

RAMOS, E. M. de F. **Toys and games in the teaching of physics.** 1990. (Master's Degree in Science Teaching). Institute of Physics and Faculty of Education, University of São Paulo.

RODITI, I. **Dicionàrio Houaiss de fisica.** Rio de Janeiro: Objetiva, 2005.

SAMPAIO, J; CALÇADA, C. **Fisica, 2ª ed,** Sào Paulo: Atual, 2005.

TIPLER, P; Mosca, G. **Fisica para cientistas e engenheiros, vol. 2, 6a ed,**

Rio de Janeiro: LTC, 2009.

YOUNG, H; Freedman, R. **Fisica III: Eletromagnetismo, 12a ed,** Sâo Paulo: Pearson, 2010.

FUNDAMENTAL BIBLIOGRAPHY

BENJAMIN, C. **Dicionário de biografias cientificas, vol.1**, Rio de Janeiro: Contraponto, 2007.

DIAS, V. S. Michael *Faraday: subsidies for experimental work methodology.* 2004. 156p. Dissertation (Master's Degree in Science Teaching) - Institute of Physics, University of Sâo Paulo

FREIRE, P. R. N. **Professora sim tia nâo - cartas a quem ousa ensinar.** Sâo Paulo, Olho D' âgua, 1997.

SANTOS, E; PIASSI,   L. P. C; FERREIRA, N. C. Activities

low-cost experiments as a strategy for building physics teachers' autonomy: an experience in continuing education. In: IX ENCONTRO NACIONAL DE PESQUISA EM ENSINO DE FÌSICA, 2004, Jaboticatubas. Available at: <http://www.cienciamao.usp.br/dados/epef/_atividadesexperimentai sd.trabalho.pdf>. Accessed on 27 Mar. 2017 20:01:34.

SERÈ, M. G; Coelho, S. M; Nunes, A. D. **O papel da experimentaçao no ensino de fisica.** Caderno Catarinense de Ensino de Fisica, Florianópolis, v.20, n.1 p. 30-42, abr. 2003.

WEBSITES CONSULTED

Image of the triboelectric series table.

Available at:

<http://lh6.ggpht.com/-

wVJFtd5Y4WA/USQ3le5o__I/AAAAAAAAInw/g5yp3dTn8kQ/clip_im

age004_thumb%25255B3%25255D.gif?imgmax=800>.

Accessed on 27 Mar. 2017 20:03:21.

Parts of Faraday's biography.

Available at:

<http://www.ghtc.usp.br/Biografias/Faraday/Faraday3.htm>.

Accessed on 27 Mar. 2017 20:05:33.

Faraday's image.

Available at:

<http://www.ghtc.usp.br/Biografias/Faraday/faradayjovem.jpg> Accessed on
27 Mar. 2017 20:06:55.

APPENDIX A - VECTOR CONSTRUCTION ELECTROSTATIC

To build the vector, you will need:

- Cardboard (approximately 5 cm);[2]

- Toilet paper;

- A canudo   ;

- A bendable straw;

- A single   ;

- Fita crepe   ,

- Base made of plaster on a disposable coffee cup with a bracket through it. The straw, rubbed with toilet paper, is placed on this bracket.

Draw an arrow on the piece of cardboard and cut it out. This arrow will play the role of the vector.

Then attach the *nylon* string behind the vector, balancing it so that it is horizontal (find the center of mass of the vector).

Stretch out the other side of the *nylon* thread, twist it around the end of the bendable straw and secure it with masking tape. Place this straw with the vector on the bracket attached to the plaster base.

Figure 12 shows a schematic drawing of the electrostatic vector. Figure 13 shows a photo of the experimental construction.

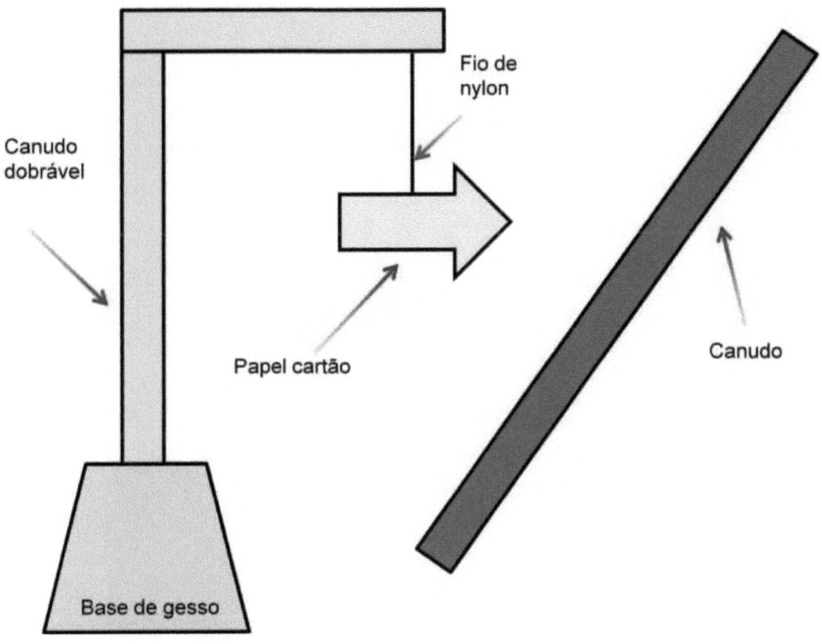

Fig. 12 - Materials used to build the electrostatic vector.

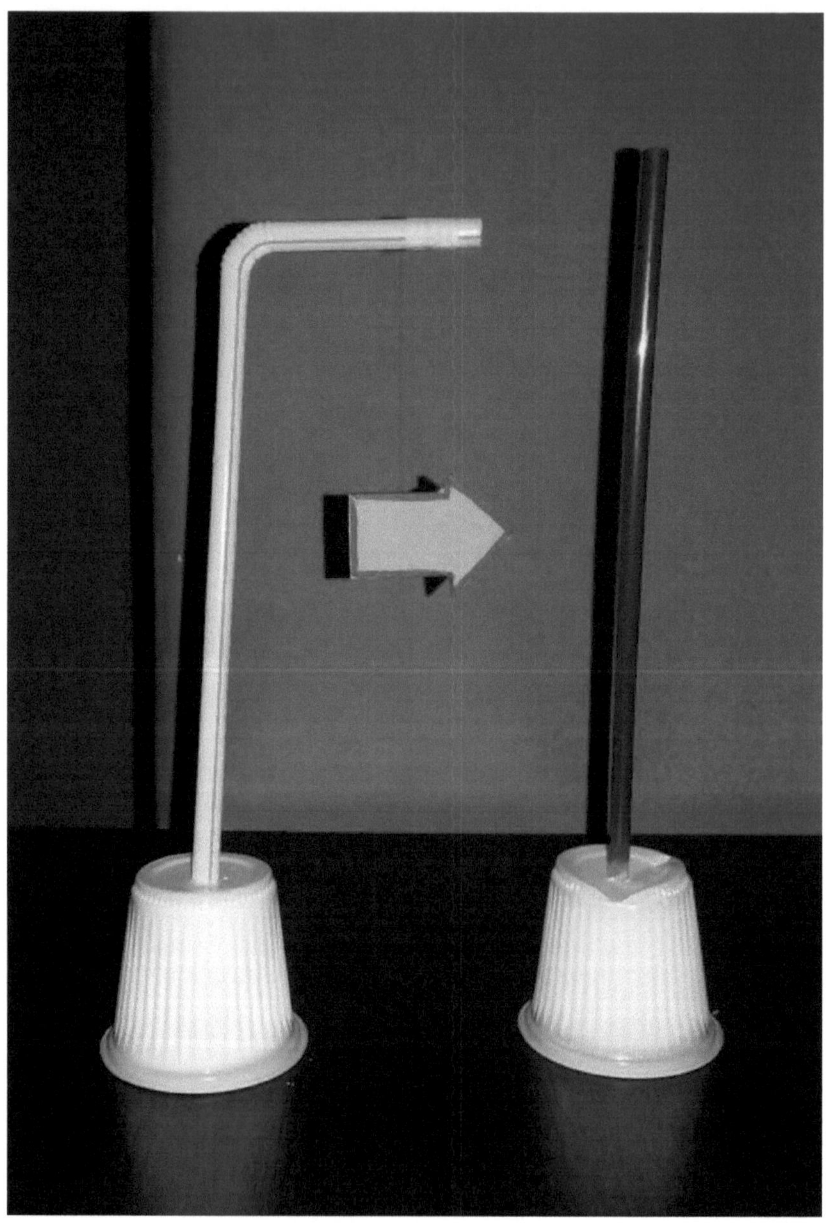

Fig.13- Photo of the Electrostatic Vector.

ACKNOWLEDGMENTS

I thank God for everything he has given me and for never letting me give up, offering me hope, patience and a lot of faith.

I would like to thank my parents, Gilson and Eleni, for sparing no effort in investing in me without charging me anything - doing everything out of love and not out of obligation and providing moral support at every moment of my life.

To my grandparents, Valter, Maria de Lourdes and Carolina, who have always supported me at every stage of my life and provided the necessary encouragement.

To my aunt and uncle Marcos and Elaine, for their conversations and admiration.

To my brother Alessandro and my cousin Luan, who taught me in practice the power of didactics.

To my little cousin Ruan, who already knows what a vector is.

To my fiancé Paulo Eduardo, who has always been by my side in good times and bad, always having a word of encouragement, for his help and guidance when needed.

To everyone who has passed through my life and contributed to who I am today and the work I do, leaving a little bit of themselves and taking a little bit of me with them.